CW00550052

AUDIO & RADIO

John Hawkins and Susan Meredith

Contents

2   The world of audio

4   How radio works

6   Making a radio programme

8   What are radio waves?

10 Transmitting sound

12 Inside a radio

14 How cassettes and records are made

18 Audio equipment

20 How audio equipment works

22 The controls on audio equipment

24 Make your own radio

27 CB and two-way radio

28 Different uses of radio and audio

30 The future

30 Audio and radio words

32 Index

Illustrated by Jeremy Banks,

Graham Round, Graham Smith, Rex Archer, Joe McEwan, Mike Roffe,
Philip Schramm and Martin Salisbury

Designed by Round Designs and Roger Priddy

The world of audio

Audio is really just another word for sound, and audio equipment is anything which reproduces sound so that we hear it through loudspeakers or headphones.

First of all, the sound has to be converted into electricity so that it can be stored and then reproduced in a different place from where it was made.

A radio is a specially complicated piece of audio equipment. It reproduces sound which has travelled as electricity, not only down wires but also through the air, on invisible waves. The sound which comes out of a radio can be either live (being made as you hear it) or pre-recorded.

Audio equipment is often called hi-fi. Hi-fi is short for "high fidelity", which means the equipment is designed to reproduce the original sound as exactly as possible.

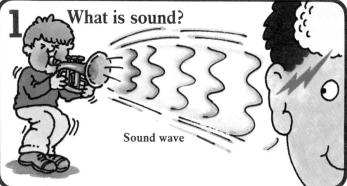

1 What is sound?

Sound wave

When someone makes a noise, tiny particles of air nearby vibrate backwards and forwards. These particles make your eardrums vibrate and produce tiny pulses of electricity, which travel along nerves to your brain. Your brain recognizes the pulses as sound. The vibrating air particles are called sound waves. It is important to remember that these are different from the waves used for radio.

2 Hands in line with speaker.

You can sometimes feel sound waves as well as hear them. A loud noise, such as thunder, can make the air vibrate so much that things shake. To feel the sound coming from a loudspeaker, try blowing up a balloon until it is firm, then hold it lightly between your hands within a metre of the speaker. Have your hands in line with the speaker. You may need to move the balloon about a bit to feel the vibrations.

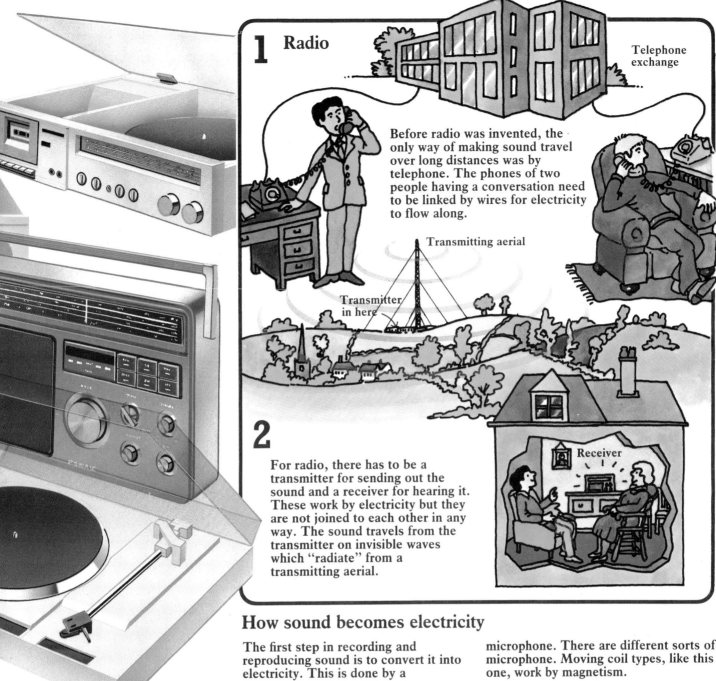

1 Radio

Telephone exchange

Before radio was invented, the only way of making sound travel over long distances was by telephone. The phones of two people having a conversation need to be linked by wires for electricity to flow along.

Transmitting aerial

Transmitter in here

2

For radio, there has to be a transmitter for sending out the sound and a receiver for hearing it. These work by electricity but they are not joined to each other in any way. The sound travels from the transmitter on invisible waves which "radiate" from a transmitting aerial.

Receiver

How sound becomes electricity

The first step in recording and reproducing sound is to convert it into electricity. This is done by a microphone. There are different sorts of microphone. Moving coil types, like this one, work by magnetism.

1 Lattice-work lets sound waves through.

2 When sound waves hit this flat disc, or diaphragm, they make it vibrate. The diaphragm vibrates at the same speed, or frequency, as the sound waves.

3 This coil of wire is attached to the diaphragm. When the diaphragm vibrates, so does the coil.

4 This is a magnet. Whenever a coil of wire moves near a magnet, an electric current is produced. This is what happens in the microphone.

5 The electric current is called a sound signal. It varies according to the frequency and loudness of the sounds. It goes down a wire connected to the coil.

3

A high note, say from a recorder, makes the air vibrate faster than a low note, say from a tuba. So, high-pitched noises are said to have a higher "frequency" than low ones. We can only hear sounds within a certain frequency range. Animals often hear things we cannot.

How radio works

These two pages will give you a general idea of the way radio works. They explain how a radio programme gets from the studio of a large radio station to your radio at home. Later in the book you can find out more about each stage in the process and about the different ways radio is used.

The studio

The people taking part in the programme talk into microphones in a studio. The sounds of their voices are converted by the microphones into electric signals. These sound signals travel through wires to the studio control room, where they are strengthened, or amplified, in the control desk. Studio engineers listen to the sounds from the studio through headphones or loudspeakers. They also have meters on the desk, which show them the sound levels. They control the volume and tone of the sounds by moving knobs and switches.

Studio

Control desk

Tape recorder

Studio control room

Programmes from other studios.

Signals from studio to continuity desk.

The continuity desk

◀ The signals then go through wires to the continuity desk. The people here have the job of linking programmes together in the right order. They select signals first from one studio, then from another. These include signals coming through wires from tape recorders, when programmes are not being broadcast live but have been pre-recorded in the studio. Continuity announcers do the talking between programmes and insert things like time-signals and jingles.

The control room

From the continuity desk, the ▶ signals go to the control room. Large radio stations broadcast several programmes at the same time on different channels. Each channel has its own continuity desk and the control room has to sort out the signals from each one. From the control room, engineers send the signals out to transmitters, usually through cables similar to those used for telephones.

Signals from continuity desk to control room.

Control room

Signals from other continuity desks.

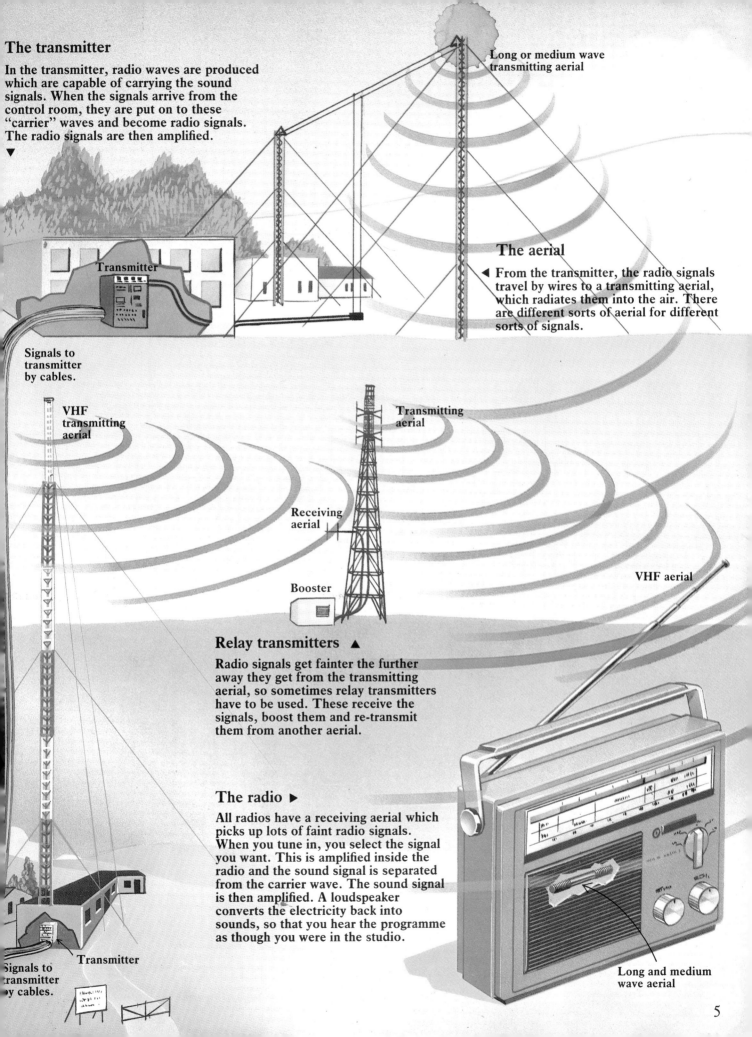

The transmitter

In the transmitter, radio waves are produced which are capable of carrying the sound signals. When the signals arrive from the control room, they are put on to these "carrier" waves and become radio signals. The radio signals are then amplified.

▼

Transmitter

Long or medium wave transmitting aerial

The aerial

◀ From the transmitter, the radio signals travel by wires to a transmitting aerial, which radiates them into the air. There are different sorts of aerial for different sorts of signals.

Signals to transmitter by cables.

VHF transmitting aerial

Transmitting aerial

Receiving aerial

Booster

VHF aerial

Relay transmitters ▲

Radio signals get fainter the further away they get from the transmitting aerial, so sometimes relay transmitters have to be used. These receive the signals, boost them and re-transmit them from another aerial.

The radio ▶

All radios have a receiving aerial which picks up lots of faint radio signals. When you tune in, you select the signal you want. This is amplified inside the radio and the sound signal is separated from the carrier wave. The sound signal is then amplified. A loudspeaker converts the electricity back into sounds, so that you hear the programme as though you were in the studio.

Signals to transmitter by cables.

Transmitter

Long and medium wave aerial

5

Making a radio programme

Some radio programmes are fairly straightforward to organize. For discussions, chat shows, panel games and even plays, the people taking part can sit round a table in a studio, talking into microphones. A news and record programme, like the one shown in these pictures, is more complicated. So many different bits have to be incorporated, including pre-recorded snippets and items produced outside the studio, that split-second timing and co-ordination are essential.

Soundproof window between studio and control room.

Control room

If the music is on tape, it is sent out by the engineer at the right time.

Very accurate clocks help the DJ time the programme exactly.

This is a "phone selector", used for phone-in programmes.

Faders

Faders control volume.

Preparing a news and record programme

The producer, disc jockey, engineer and production secretary meet about a week before the programme to plan any special items, such as interviews with famous people. The day before the programme, they finalize details such as which records to play, which dedications to read out and the exact timing of the programme.

The news

Sometimes the DJ reads the news, sometimes it goes from a separate studio direct to the engineer's desk. The DJ plays a jingle to announce the news and the engineer "fades up" a channel on the desk, linked to the newsroom.

On-the-spot reporters

On-the-spot reporters do interviews using high quality, portable recorders. The tapes are taken back to the studio and sent out from the control room. They can be "edited" to make them the right length.

DJs can talk to the listeners or the engineer through their microphones, depending on how they set their microphone switch. For "voice-overs", the music is automatically faded down whenever the DJ speaks.

After going through the engineer's desk, the signals leave the control room through cables in this "jackfield". Items from outside the studio arrive at the desks through the jackfield.

The engineer feeds the sound back from the control desk to the DJ's headphones and can talk to the DJ through a microphone.

Engineer's control desk

Delay box for phone-in programmes.

DJ's control desk

Pre-recorded cassettes of jingles, adverts or sound effects are played in this "jingle box".

There are two turntables. While one record is playing, the DJ gets the next one ready on the other turntable, so it can be switched on as soon as the first record ends.

In the DJ's studio

Disc jockeys are responsible for linking the different parts of their programme smoothly together and for pacing the programme correctly. They have a rough script, agreed with the producer beforehand, so they know what order to do everything in, but they also improvise a lot. The signals go from the DJ's control desk to the engineer's desk. Engineers make sure the signals go to the transmitter without any technical hitches. They control the volume and tone of the sounds. Items from outside the studio also go through the engineer's desk before going to the transmitter.

Phone-ins

When listeners phone in, the producer decides whose calls to include. The production secretary phones the callers back when they are wanted on the air. The calls reach the DJ through the "phone selector", which is like a small switchboard. The microphone and headphones act as the DJ's telephone. The callers' words are delayed by about a second so the engineer can cut out anything abusive.

Remote studios

Signals go from one studio to other by cables.

Speakers convert signals from studio into sound.

Engineer

Remote studios are used for interviews at places like airports. The signals have to go through the control desks in the usual way, so the remote studio is linked to the main studio by wires. The link is two-way, so the people being interviewed can hear the DJ. The control room engineer and the remote studio engineer can communicate with each other by telephone.

Radio cars

Mast has two aerials at top, one for interview, the other for "off-air" communication.

Microphone and speaker for "off-air" communication.

Interviewer listens to instructions from studio on headphones.

Transmitters

These are linked to the studio by radio so they can move around. There are two separate radio links: one for the interview and one for "off-air" communication between the studio and the interviewer. The interviewer often drives the car and acts as an engineer as well.

What are radio waves?

Radio waves are pulses of electrical energy, which can travel through air, space and even solid objects. You will sometimes hear them referred to as electromagnetic waves. Some radio waves come from stars or are created by lightning but these cannot be used to carry sound. To carry sound, the waves have to be made in a transmitter.

The term "radio waves" sometimes means the waves made in the transmitter, before they have any sound on them. (These are also called carrier waves.) Or it can mean the waves which go out from the aerial carrying sound.

When you throw a stone into a pond, circles of ripples spread outwards from it. You can imagine radio waves looking and moving like these ripples.

Radio waves usually travel from transmitting aerials in circles, like the ripples on the pond, but they can also be made to travel in beams.

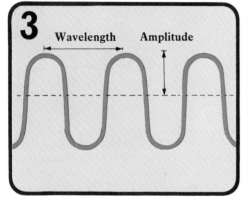

The wavelength of waves used for radio varies between 0.33mm and 30km. The distance is measured from the crest of one wave to the crest of the next. The height of a wave is called its amplitude and indicates its strength.

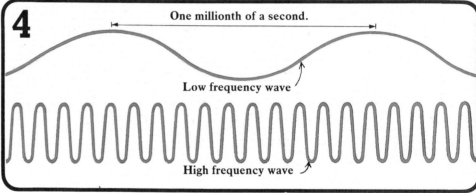

All radio waves travel at the speed of light (300,000km per second). This means that the shorter the wavelength the greater the rate, or frequency, at which complete waves go past a certain point. Frequency is measured as the number of times every second a wave goes by. (Because they go so fast, the diagram above shows waves going by in a millionth of a second.) Different frequencies are used for different sorts of radio transmission.

1 Make your own radio waves

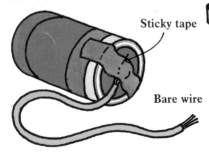

You can make and transmit radio waves to your radio. Using a sharp knife, carefully strip about 2cm of the plastic off both ends of a piece of electrical wire about 40cm long. Fasten one end firmly to either end of a cylindrical shaped battery with sticky tape.

2

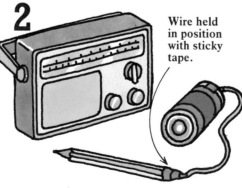

Sharpen a pencil at both ends. Twist the loose end of wire neatly round one of the points and fasten it firmly with sticky tape. Tune a portable radio to part of the long or medium waveband where there is no programme and put the battery near it.

3

Moving the pencil and battery round the radio, tap the free pencil point against the free end of the battery. You will hear clicks on the radio as it picks up the radio waves you are making. If you move away from the radio, the signals will get fainter.

How radio waves were discovered

The first person to prove that he had produced radio waves was the German scientist, Heinrich Hertz, in 1887. He set up two metal spheres with a small gap between them and, at the other side of his laboratory, a wire loop with a gap in it. When he used an electric current from a battery to make a spark jump the gap between the metal spheres, he noticed that a spark appeared immediately in the gap in the wire loop. He realized that he had created and transmitted a radio wave from the spheres to the loop.

Radio waves similar to the ones produced by Hertz are made whenever the electric current in a wire changes suddenly, for instance, when electrical equipment is switched on or off. Sometimes the waves show up as interference on a nearby radio or TV.

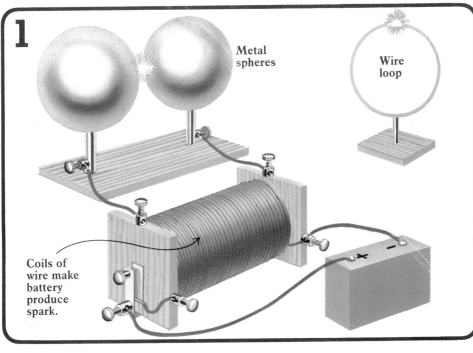

Metal spheres

Wire loop

Coils of wire make battery produce spark.

Different types of waves

Radio waves are part of a family of waves known as the electromagnetic spectrum. This includes various light and heat waves. The only difference between the waves is their length. Radio waves are the longest. None of our senses can detect them. We feel shorter, infra-red waves as heat. At shorter lengths still, our eyes detect the waves as light of different colours. And at even shorter lengths, there is ultra-violet light, which is used in sun-ray lamps. X-rays and gamma rays are also part of the electromagnetic spectrum. Sound waves are not.

Radio waves go through our bodies without our knowing.

Infra-red rays can be used for cooking.

Visible light is reflected into our eyes to make us see.

Ultra-violet light can be used to produce a sun-tan.

Transmitting sound

To carry speech or music, radio waves have to be continuous, not like the ones made by early radio scientists such as Hertz, which only occurred in short bursts. Continuous carrier waves are made, or generated, in radio transmitters. The waves then have the sound signals put on to them in a process called modulation. You can find out about the two different sorts of modulation below. After modulation, the waves are fed through wires to an aerial and from the aerial they are radiated into the air for your radio to pick up.

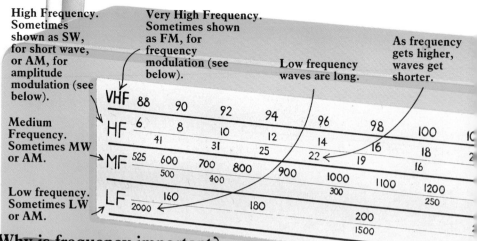

High Frequency. Sometimes shown as SW, for short wave, or AM, for amplitude modulation (see below).

Very High Frequency. Sometimes shown as FM, for frequency modulation (see below).

Low frequency waves are long.

As frequency gets higher, waves get shorter.

Medium Frequency. Sometimes MW or AM.

Low frequency. Sometimes LW or AM.

VHF	88	90	92	94	96	98	100	
HF	6	8	10	12	14	16	18	
	41	31	25	22	19	16		
MF	525 600 700 800		900	1000	1100	1200	
	500 400			300		250	
LF	160	180		200			
	2000		1500				

Why is frequency important?

Radio waves have to have different lengths and frequencies, or they would get mixed up together and it would not be possible to hear one broadcast separately from another. Certain frequencies are reserved by international agreement for use by different radio stations. Other frequencies are kept for other users.

You can see from the radio tuning display above how the waves are divided into bands. This radio has low, medium, high and very high frequency bands. On some displays, low, medium and high frequency waves are referred to by their length (long, medium and short wave). Wavelength is measured in metres. Frequency is measured in kilohertz

1 Putting sound on the waves

Continuous carrier wave.

Strong sound signal   No sound signal   Weak sound signal

Signal from microphone.

Amplitude modulated wave.

One way of putting sound on radio waves (modulating them) is to alter their amplitude (height). When the signal from the microphone is strong, the amplitude varies a lot. When the signal is weak, the amplitude varies less. This is called amplitude modulation (AM). Low, medium and high frequency waves are amplitude modulated.

2

Continuous carrier wave.

Strong sound signal   No sound signal   Weak sound signal

Signal from microphone.

Frequency modulated wave.

The other way of putting sound on the waves is called frequency modulation (FM). When the signal from the microphone is strong, the frequency varies a lot. When the signal from the microphone is weak, the frequency varies less. The amplitude of the waves stays the same all the time. VHF waves are frequency modulated.

1 How the waves travel

Long and medium waves keep fairly close to the Earth, following its shape. Long waves can travel for over a thousand kilometres, medium waves for several hundred, before they lose their energy and eventually fade away.

2

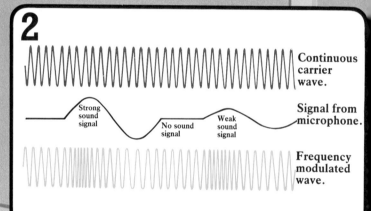

Short waves go up to a layer in the Earth's upper atmosphere called the ionosphere. This reflects the waves back to Earth, so they travel a very long way. For this reason, short waves are used for overseas broadcasting.

3

Very high and ultra-high frequency waves do not go through solid objects very well, so they are used for broadcasting over short distances (up to about 150km). They also go off into space right through the ionosphere.

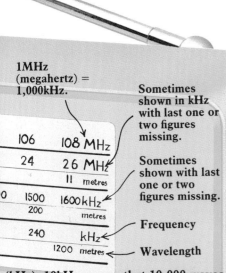

1MHz (megahertz) = 1,000kHz.

Sometimes shown in kHz with last one or two figures missing.

106	108 MHz
24	26 MHz
	11 metres

Sometimes shown with last one or two figures missing.

1500	1600 kHz
200	metres
240	kHz
1200 metres	

Frequency

Wavelength

(kHz). 10kHz means that 10,000 waves go past a certain point in a second. The frequency of waves used for broadcasting varies from 10kHz to about 12 million kHz, or 12,000 megahertz (MHz). Sometimes, if there is not much space on the display, the last figure or two of the kHz is missed off.

3

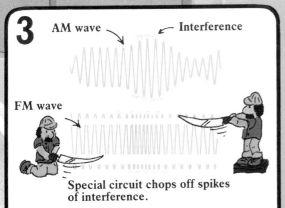

AM wave — Interference

FM wave

Special circuit chops off spikes of interference.

Amplitude modulated waves often get "spikes" of interference on them. These are caused by people using electrical equipment and are heard as clicks and crackles on the radio. Frequency modulation cuts out interference but long and medium waves cannot usually be frequency modulated.

How a transmitter works

A radio transmitter has three main parts. An "oscillator" generates the carrier waves. A "modulator" puts the sound signals on them. An "amplifier" boosts the modulated waves. All three parts consist of lots of electronic components arranged in circuits. Here is a simplified picture of an AM transmitter.

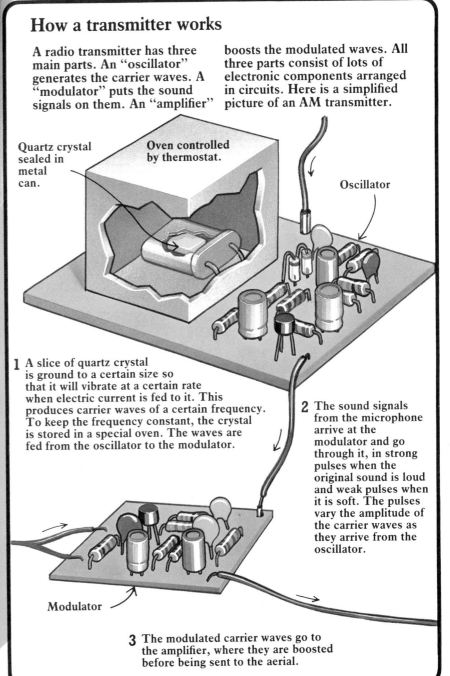

Quartz crystal sealed in metal can.

Oven controlled by thermostat.

Oscillator

Modulator

1 A slice of quartz crystal is ground to a certain size so that it will vibrate at a certain rate when electric current is fed to it. This produces carrier waves of a certain frequency. To keep the frequency constant, the crystal is stored in a special oven. The waves are fed from the oscillator to the modulator.

2 The sound signals from the microphone arrive at the modulator and go through it, in strong pulses when the original sound is loud and weak pulses when it is soft. The pulses vary the amplitude of the carrier waves as they arrive from the oscillator.

3 The modulated carrier waves go to the amplifier, where they are boosted before being sent to the aerial.

4

Super-high frequency waves go through the ionosphere into outer space and are used for satellite broadcasting. The signals can be beamed up from one country and then sent down by the satellite to several others at once.

Getting good reception

Long and medium waves travel best over water or damp ground. If you live in an area with clay soil or lots of rivers, you will probably get good reception of programmes on these wavebands. Reception is usually worst in areas where the rocks are very old.

Picking up more stations at night

During daylight, some of the energy from medium waves goes into space but after dark it gets reflected back from the ionosphere. This means you often pick up medium wave foreign programmes at night that you cannot hear in the day, and you also get more interference.

Inside a radio

Your radio is technically called a radio "receiver", because it receives the signals sent from the transmitter. Most of the work in a radio is done by lots of electronic components like the ones in the picture below. There are usually more of each type of component than are shown here. They are arranged in circuits. On the opposite page you can find out what happens inside your radio.

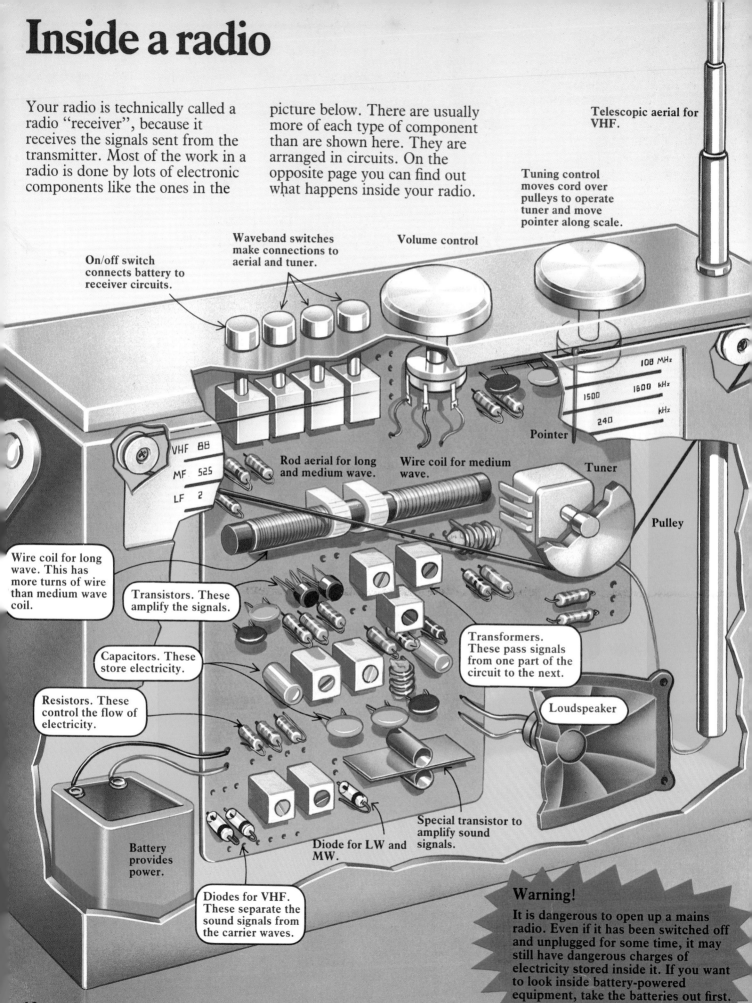

Telescopic aerial for VHF.

Tuning control moves cord over pulleys to operate tuner and move pointer along scale.

On/off switch connects battery to receiver circuits.

Waveband switches make connections to aerial and tuner.

Volume control

108 MHz
1500   1600 kHz
240   kHz

Pointer

Rod aerial for long and medium wave.

Wire coil for medium wave.

Tuner

Pulley

VHF 88
MF 525
LF 2

Wire coil for long wave. This has more turns of wire than medium wave coil.

Transistors. These amplify the signals.

Capacitors. These store electricity.

Resistors. These control the flow of electricity.

Transformers. These pass signals from one part of the circuit to the next.

Loudspeaker

Battery provides power.

Diode for LW and MW.

Special transistor to amplify sound signals.

Diodes for VHF. These separate the sound signals from the carrier waves.

Warning!

It is dangerous to open up a mains radio. Even if it has been switched off and unplugged for some time, it may still have dangerous charges of electricity stored inside it. If you want to look inside battery-powered equipment, take the batteries out first.

How your radio works

Your radio has to pick up the waves coming from the transmitting aerial and separate the sound signals from the carrier waves so they can be converted back into sound that you can hear. When you look inside a radio, it is hard to imagine the different stages in this process. The balloons and robots in this picture should help you understand what happens to the signals on their way from the radio aerial to the loudspeaker. The balloons are carrier waves and the notes are sound signals.

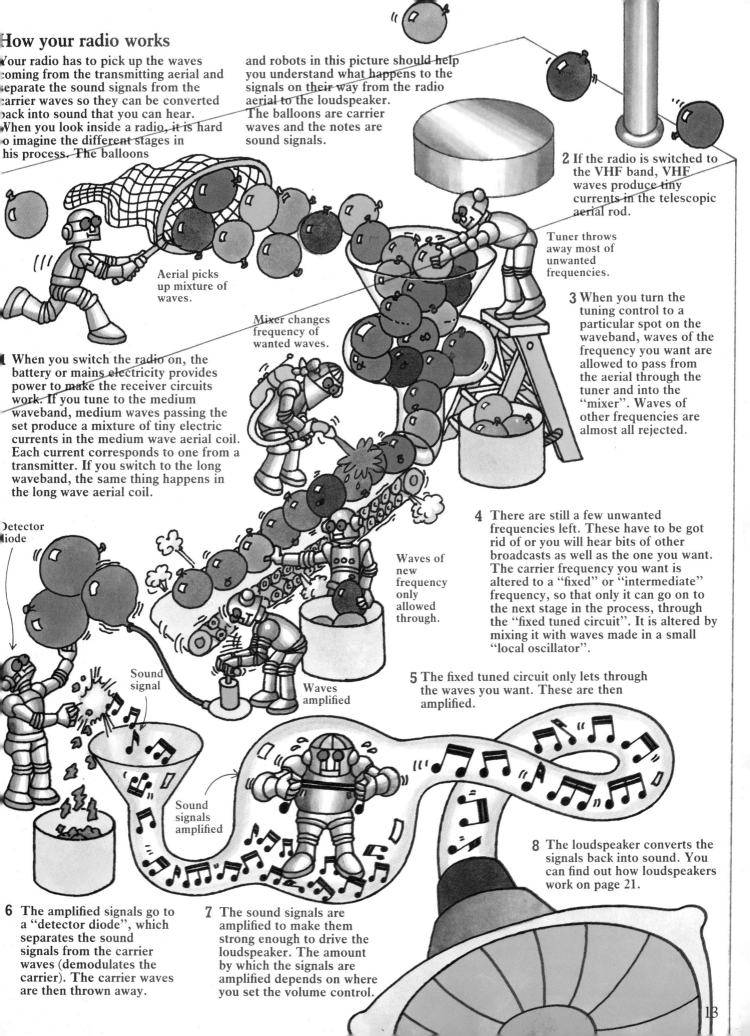

Aerial picks up mixture of waves.

Mixer changes frequency of wanted waves.

Detector diode

Sound signal

Sound signals amplified

Waves amplified

Waves of new frequency only allowed through.

1 When you switch the radio on, the battery or mains electricity provides power to make the receiver circuits work. If you tune to the medium waveband, medium waves passing the set produce a mixture of tiny electric currents in the medium wave aerial coil. Each current corresponds to one from a transmitter. If you switch to the long waveband, the same thing happens in the long wave aerial coil.

2 If the radio is switched to the VHF band, VHF waves produce tiny currents in the telescopic aerial rod.

Tuner throws away most of unwanted frequencies.

3 When you turn the tuning control to a particular spot on the waveband, waves of the frequency you want are allowed to pass from the aerial through the tuner and into the "mixer". Waves of other frequencies are almost all rejected.

4 There are still a few unwanted frequencies left. These have to be got rid of or you will hear bits of other broadcasts as well as the one you want. The carrier frequency you want is altered to a "fixed" or "intermediate" frequency, so that only it can go on to the next stage in the process, through the "fixed tuned circuit". It is altered by mixing it with waves made in a small "local oscillator".

5 The fixed tuned circuit only lets through the waves you want. These are then amplified.

6 The amplified signals go to a "detector diode", which separates the sound signals from the carrier waves (demodulates the carrier). The carrier waves are then thrown away.

7 The sound signals are amplified to make them strong enough to drive the loudspeaker. The amount by which the signals are amplified depends on where you set the volume control.

8 The loudspeaker converts the signals back into sound. You can find out how loudspeakers work on page 21.

How cassettes and records are made

On the next four pages you can find out how pop cassettes and records are made. Both cassettes and records are usually made from a series of tapes, which are recorded in a studio. Tape is used because it can be corrected, or edited, fairly easily.

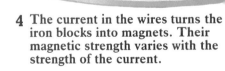

The studio

The musicians play into microphones in a soundproofed studio. They are separated from each other by screens, so the sounds from one instrument are not picked up by anyone else's microphone. The microphones convert the sounds into electric signals and these go down wires to a control desk, or mixer, in the control room. The musicians wear headphones so they can listen to the overall sound, which is fed back to them from the control room.

How sound gets on to tape

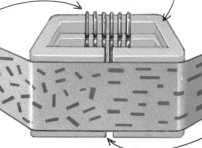

1 The part of a tape recorder which puts the sound on tape is called the recording head. The unrecorded tape is fed to the head from this feed spool.

2 The tape is made of plastic. It is coated with millions of tiny, invisible particles of a substance called iron oxide.

3 The sound signals come from the microphones, through the desk, to these coils of wire in the recording head. Each coil of wire is wrapped round an iron block. There has to be a separate coil and block for each track on the tape (see opposite page). The current in the wires varies according to the sounds.

4 The current in the wires turns the iron blocks into magnets. Their magnetic strength varies with the strength of the current.

The desk

The desk has lots of identical channels, or modules, one for each microphone. The engineer can control the signals from each microphone separately. Here is a typical module.

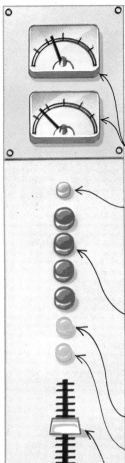

The VU meters measure volume units and show the engineer how loud the sound is.

This knob cuts out an instrument if it is not needed for part of the piece. Or it can "solo" the instrument for the engineer to check, by lowering the sound levels of all the other instruments.

These frequency knobs alter the tone of the sounds. For example, if a bass guitar sounds too booming, its low notes can be made softer and its high notes louder.

Echo knob (see page 16).

Stereo pan pot (see page 16).

This "fader" slides up and down to alter the volume of the sounds.

The control room

The producer and sound engineer listen to the sound in the control room, through loudspeakers or headphones. The engineer amplifies the sound signals by moving knobs and switches on the desk. The producer can give instructions to the musicians by talking into a microphone. The musicians hear the instructions on their headphones.

6 The tape now has sound stored on it in a magnetic pattern. It goes to the take-up spool. (You can find out how the sound is played back on page 21.)

As the tape passes across a tiny gap at the front of each coil and block, the particles of iron oxide on the tape are magnetized by the blocks. Their magnetic strength varies with the strength of the current and the magnetic strength of the blocks.

Correcting tape

The tape can be fed past an "erase head" in the recorder. A strong electric current demagnetizes the tape, removing the sound so a new recording can be made. Tape of certain widths is also edited by having sections cut out and the ends rejoined.

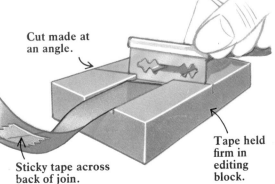

Cut made at an angle.

Sticky tape across back of join.

Tape held firm in editing block.

The recorder

The signals do not go to the tape recorder until the producer is satisfied with the sound. Then they are sent down wires from the desk. The first recording is usually made on tape which is two inches wide. Up to 24 tracks are recorded running parallel to each other across the tape. Sometimes each microphone has its own track. Sometimes the signals from two microphones are combined. The tracks do not have to be recorded at the same time, so the musicians do not all have to be at the recording session together. If someone makes a mistake, their track can be re-recorded separately. This is called overdubbing.

The master tape

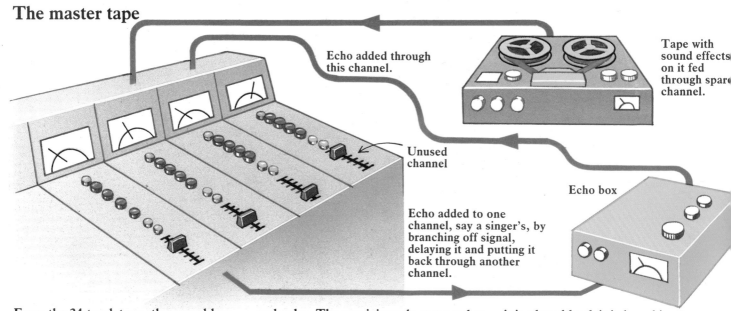

Echo added through this channel.

Tape with sound effects on it fed through spare channel.

Unused channel

Echo box

Echo added to one channel, say a singer's, by branching off signal, delaying it and putting it back through another channel.

From the 24-track tape, the sound has to be "mixed down". This is done by feeding the signals back through the desk again from the recorder. The engineer makes alterations to the tone and can add things like sound effects and echo. The musicians do not need to be there. In this diagram you can see how the last few modules of a desk might be used in the mix down.

The sounds are usually put on to a two-track tape eventually, so that when it is played back it is heard in stereo. (The sound from one track comes out of the left speaker, the sound from the other out of the right.) This two-track, stereo tape is ¼in wide and is called the master tape.

Producing stereo sound

When engineers are mixing down, they do not just put half of the 24 tracks on to one track of the stereo tape and the other half on the other track. That would mean the sound of half the instruments came out of one speaker and half out of the other. By using the pan pot, the engineer can make it sound as though the musicians are arranged on stage at a live concert. This diagram shows what happens if the pan pot knob on a module is moved towards the left. If the knob is moved towards the right, the opposite happens.

Pan pot knob on module moved towards left.

Only a small amount of signal goes on to right channel and comes out of right speaker.

Most of signal goes on to left channel of stereo tape.

Sound from left channel comes out of left speaker and instrument sounds as though it is on left of stage.

Making echoes

Signal split in two.

One part of signal chopped up.

Quick route

Bucket brigade delay line

Main signal with echo on top.

Delayed signal added to main signal.

One way of giving a track an echo is to delay part of its signal in a "bucket brigade delay line". Part of the signal is "chopped up" and passed on from one bit of the electronic circuit to the next in pieces. This delays it. At the end of the line, the pieces are put together again and added to the main signal as echo. This is also known as "digital delay".

16

Cutting a record

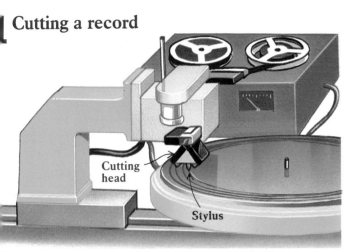

Cutting head

Stylus

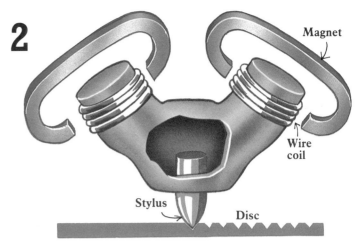

Magnet

Wire coil

Stylus

Disc

The master tape is played back on a machine which feeds the signals to a "cutting head". The head has a stylus made of diamond. This cuts a groove in a soft plastic disc which is rotated on a turntable. It starts at the outside of the disc and spirals in to the centre.

The cutting head has two coils of wire attached to the stylus, one for each track of the tape. When the electric signals from the tape reach the coils, the coils vibrate because they are near magnets. This makes the stylus vibrate. The rate of vibrations varies with the current in the coils.

3

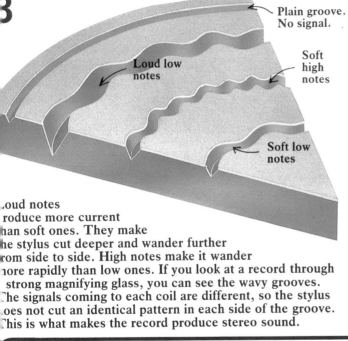

Plain groove. No signal.

Loud low notes

Soft high notes

Soft low notes

Loud notes produce more current than soft ones. They make the stylus cut deeper and wander further from side to side. High notes make it wander more rapidly than low ones. If you look at a record through a strong magnifying glass, you can see the wavy grooves. The signals coming to each coil are different, so the stylus does not cut an identical pattern in each side of the groove. This is what makes the record produce stereo sound.

4

A strong metal "stamper" is made from the disc. It has ridges in place of the grooves. Two stampers (one for each side of the record) are put face to face in a press. They have the record labels on them upside down. One stamper has a ball of PVC material on it. The stampers are heated and brought together. The ridges press into the PVC, "printing" both sides of the record. The record is cooled, trimmed and put in its sleeve.

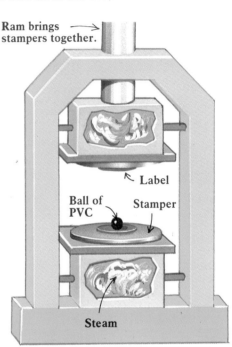

Ram brings stampers together.

Label

Ball of PVC

Stamper

Steam

Producing cassettes

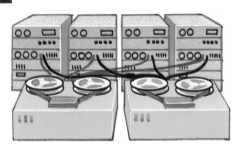

Cassettes are easier to produce than records. Two master tapes (one for each side of the cassette) are played back at the same time on machines feeding lots of cassette recorders at once. The tapes are speeded up. A cassette which plays for 90 minutes is recorded in 2½ minutes.

2

First side of cassette (tracks 1 and 2) is played to end.

Top of cassette

Track 4
Track 3
Track 2
Track 1

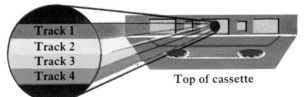

Track 1
Track 2
Track 3
Track 4

Top of cassette

Cassette is turned over for tracks 3 and 4, and starts at end where it just finished. Sound would be back to front if master tape had not been played backwards when recording.

Tape can only be recorded on one face. Cassette tape is ⅛in wide and has four tracks on it. This gives you two "sides" of two-track, stereo sound. One of the master tapes has to be played backwards for the sound to come out the right way round when you turn the cassette over.

Audio equipment

These two pages show you some of the different types of audio equipment that are available and how they can be combined. If you are thinking of buying equipment, it is a good idea to plan ahead, as you might be able to build up a good system in stages. Always get any item of equipment demonstrated in the shop before you buy. Remember that the word "hi-fi" is often used instead of "audio" but it really only applies to top quality equipment.

On the right is a typical home hi-fi system of "separates". If you want to build up a system of separate units made by different manufacturers, check with them or the supplier that the units will all work together.

▲
If you want to play records, you will need a turntable. You can get turntables with built-in amplifiers, but these are getting rare and are not usually very good quality.

You can use cassette decks for playing tapes and recording, either from another piece of hi-fi equipment or live from a microphone.

Rack systems

These have a separate ▶ turntable, cassette deck, tuner and amplifier made by the same manufacturer and stacked in a rack. There are speakers to match. Most rack systems are good quality and fairly expensive. You can buy the units individually and build up the system gradually.

Music centres

◀ These have a turntable, cassette deck, tuner and amplifier all in one unit and just need speakers adding. They are usually cheaper and not always as good as rack systems. They are convenient because they do not need much wiring.

Mini hi-fi systems

These include a cassette ▶ deck, tuner, amplifier and speakers, and sometimes a turntable, all in a space about the size of a small suitcase. The units can be positioned apart or kept together. Many are portable. Some are about as good as rack systems but they are quite expensive.

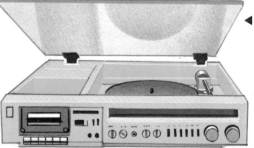

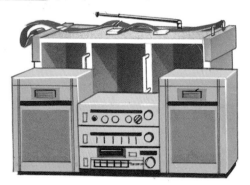

Radio/cassette recorders

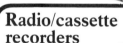

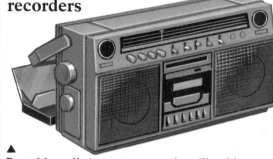

▲
Portable radio/cassette recorders like this are often quite good quality. The speakers are not far enough apart to give a very good stereo effect but many models can be connected to headphones or separate speakers. One advantage is that you can record from the radio without having to connect two machines. Many of the radios have four wavebands. There is often a socket for a separate microphone. The machines work from batteries or the mains. You can also get portable mono (one track) radio/cassette recorders, with just one speaker. These are cheaper.

Smaller stereo ▶ radio/cassette recorders have headphones instead of speakers. Most work only from batteries. There are often sockets for a second pair of headphones and a separate microphone.

18

◀ Speakers provide stereo sound. You can sometimes link up three or four speakers to your amplifier for a "surround-sound" effect.

The amplifier boosts the signals from the turntable, cassette deck or tuner before sending them to the speakers. ▼

Headphones plug into the amplifier.

◀ Tuners are radios without built-in amplifiers. They provide good quality VHF radio in stereo, with an outdoor aerial. Not all programmes are broadcast on VHF, so get a tuner that has long and medium wavebands as well.

▲ You can plug a microphone into a cassette deck to make live recordings, say of someone talking or playing a musical instrument.

Radios

Most portable radios do not have more than three wavebands and some only have two. They produce mono sound. They can be run from the mains as well as batteries. It is cheaper to use the mains when possible.

If radio is to be the most important part of your hi-fi system, you might start by buying a "receiver". This is a tuner and amplifier in one unit, and produces good stereo sound through hi-fi speakers. A turntable and cassette deck can be connected up to the amplifier part of the receiver. ▼

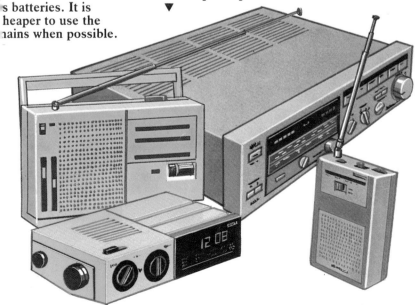

Radio alarm clocks can be set to wake you up to the sound of the radio. Most have three wavebands and produce mono sound. Some have a cassette recorder combined. They run from the mains.

▲ Most micro radios have only two wavebands. They produce mono sound and work from batteries.

Cassette recorders

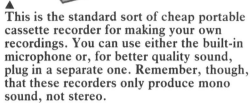

▲ This is the standard sort of cheap portable cassette recorder for making your own recordings. You can use either the built-in microphone or, for better quality sound, plug in a separate one. Remember, though, that these recorders only produce mono sound, not stereo.

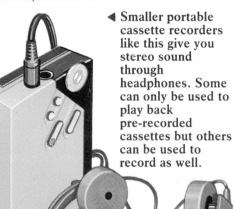

◀ Smaller portable cassette recorders like this give you stereo sound through headphones. Some can only be used to play back pre-recorded cassettes but others can be used to record as well.

Most micro ▶ cassette recorders do not give good enough quality sound for music, but are used for recording and playing back speech.

Car audio equipment

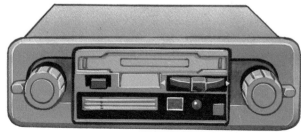

▲ For cars you can get radios and cassette players either combined or separate, often with a clock. Speakers can be fitted to the doors or the back shelf for stereo sound. There are also special mini hi-fi systems for cars.

How audio equipment works

Although there are many different types and makes of audio equipment, they all work on the same basic principle. That is, they have to convert sound which is stored on tape or disc back into electric signals and then into sound that you can hear. Here you can find out how the equipment works and pick up some tips and hints for when you are buying equipment.

Turntables

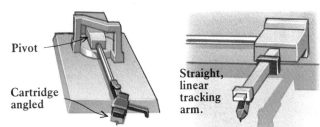

There are three main types of turntable. In the cheapest type, the motor turns a rubber idler wheel, and this makes the turntable go round. Vibrations from the motor are sometimes transferred to the turntable by the idler wheel, producing "rumble" in the speakers. In belt-driven systems, the motor is connected to the turntable by a rubber belt. This cuts down vibrations. The best and most expensive type of turntable is direct drive. Here, the turntable is mounted directly on a special, slow-turning motor.

Tonearms

Pivot

Cartridge angled

Straight, linear tracking arm.

Most tonearms have a pivot at one end and are bent at the other, near the cartridge. This holds the stylus at the correct angle. The best arms are "linear tracking". These are completely straight and are driven along a bar as the record plays. They are very expensive.

Cartridges

Magnet

Stylus vibrations transferred to magnet.

Magnets

Stylus vibrations transferred to coils.

Stylus

Currents produced in coils because they are near moving magnet.

Currents produced in coils because they are moving near magnets.

There are two main sorts of cartridge: moving magnet and moving coil. Moving coil types work on the same principle as the disc cutting cartridge on page 17. Make sure your cartridge is suitable for your amplifier.

Try to put your speakers on a level with each other and with your head. They should be between two and four metres apart. There are often two or three speakers in each speaker box to handle different pitched sounds. Make sure your speakers have the right "impedance" (see page 31) for your amplifier.

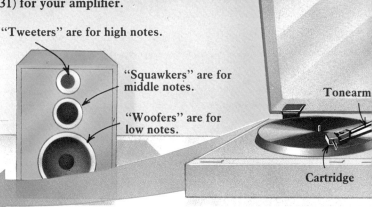

"Tweeters" are for high notes.

"Squawkers" are for middle notes.

"Woofers" are for low notes.

When you put a stylus on to a revolving record, the pattern in the sides of the record groove makes it vibrate. The vibrations are transferred to the cartridge which produces tiny electric signals. These are fed through wires in the tonearm to the amplifier. Styluses are made of sapphire or diamond. Diamond ones are best.

Tonearm

Cartridge

Headphones work like loudspeakers. Open headphones, like the pair in this picture, allow more sound to escape than closed headphones, and allow you to hear more of what is going on around you. Like speakers, headphones have to have the right impedance for the amplifier you are using.

Cassette decks

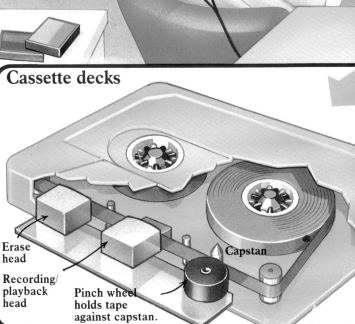

Erase head

Recording/ playback head

Pinch wheel holds tape against capstan.

Capstan

Some people think records produce better quality sound than cassettes. On the other hand, they take up more room and are more easily damaged. Digital records are better than ordinary ones. You can find out more about these on page 30. Always keep records in their sleeves and stack them side by side. Clean them with special cloths from record shops.

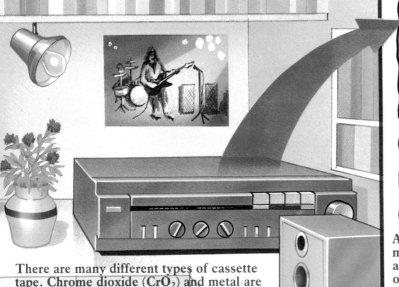

Amplifiers

Signals boosted.

Tone and balance controlled.

Signals boosted again.

All electric sound signals have to be amplified to make them strong enough to drive loudspeakers. The amplifier picks up tiny signals from whichever piece of equipment, or "source", you switch on, and boosts them by transistors. It sets the tone and balance of the sounds. Then it boosts the signals about a million times more. It is important not to buy an amplifier too powerful for the system's speakers. Sometimes, instead of being "integrated", an amplifier is split into two parts: a preamplifier and a power amplifier.

There are many different types of cassette tape. Chrome dioxide (CrO_2) and metal are the best. Cassettes are labelled C10, C30, C60, C90 or C120. The figures show the total running time (both sides) in minutes. C120s are made of thinner plastic than the others so enough tape can be squeezed into the cassette. This means they break more easily.

Keep cassettes in a cool, dry place, away from electrical equipment, which can erase recordings. This includes vacuum cleaners, TV sets and even speakers. Playing a head-cleaning cassette occasionally helps to preserve the quality of the recordings.

You will come across two main types of microphone: dynamic or moving coil (see page 3) and capacitor or condenser. These can both be either omni-directional, which means they pick up sound from all around, or uni-directional (cardioid), which means they pick up sound mainly from the front.

Loudspeakers

Electric signals come from the amplifier to a wire coil attached to the speaker cone. This electric current makes the coil vibrate, because it is near a magnet. This then makes the cone vibrate, producing sound waves in the air. Speakers work like moving coil microphones in reverse (see page 3).

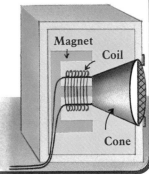

Magnet

Coil

Cone

Tape is pulled along by a capstan, driven by an electric motor. Sound is stored on tape in a magnetic pattern. When a recorded tape passes across the playback head, it magnetizes two iron blocks in the head (one for each stereo track). This produces electric currents in the wires wrapped round each block. The currents vary according to the magnetic pattern on the tape. They go to the amplifier. (To find out how sound gets on to tape when you make a recording, see page 14.)

Most cassette decks have two heads, one for playback and recording, the other for erasing. When you press the erase button, a supersonic current, produced in a special circuit, destroys the magnetic pattern on the tape, leaving it ready for a new recording.

Reel-to-reel recorders

These are expensive and are used by recording enthusiasts. There are not many pre-recorded tapes for them. They use wider tape than cassette decks (¼in) and can be played at faster speeds. This improves the quality of the sound. You can edit by cutting as well as erasing (see page 15).

The controls on audio equipment

These two pages show you what the different controls on audio equipment are for and explain some technical audio words. The controls are shown here on a rack system but many are very similar on individual or smaller, portable pieces of equipment. Not all equipment has all these controls. On pages 30-31 there are some more technical words that you will come across, particularly in makers' specifications for equipment.

SPEED SELECT. Records have to be spun at the same speed as when they were being cut. These controls make the turntable revolve at 33 r.p.m. (revolutions per minute) for LPs (long play) and 45 r.p.m. for EPs (extended play).

RECORD/PLAYBACK. You press these controls to record, play back a tape, re-wind, move the tape forward fast, pause while recording, and to stop the tape and open the door so the cassette can be taken out.

POWER INDICATORS. These show the power output of the amplifier. Here, a lot of power is going to the left speaker and not much to the right. When an amplifier's output reaches the top of its power range, the sound often gets distorted.

SOURCE OR INPUT. This control selects the signals from whichever piece of equipment you want to play. This amplifier can handle signals from a turntable with either a moving magnet or moving coil cartridge, from a tuner, or from either of two cassette decks or reel-to-reel recorders. The auxiliary position means you can link up another piece of equipment to your system, perhaps a friend's cassette deck for recording from their machine to yours, or an electronic musical instrument.

SIGNAL STRENGTH INDICATOR. This shows whether you are tuned to a station exactly and whether you are receiving signals as clearly as you should.

STROBOSCOPE. Some turntables have patterns of bars round the edge and a neon light which shines on them. When the turntable speed is correct, the bars look stationary under the light.

QUARTZ LOCK. Quartz-locked turntables have a special circuit for keeping their speed accurate. This cuts down "wow" (distorted sound caused by slow fluctuations in speed) and "flutter" (distortion caused by rapid fluctuations in speed). The phrase "phase-locked loop" (PLL) also applies to these turntables.

POWER. The on/off control supplies mains electricity to power the equipment.

TWO MOTOR. Some cassette decks have one motor to drive the capstan and another to drive the spools. This helps keep the tape speed constant and so cuts down wow and flutter (see above).

PHONES. Check that your headphones' "impedance" is suitable for the deck.

TAPE COUNTER. This shows how far along the tape you are, so you can make a note of where to find a particular track. Pressing the re-set button puts the counter to zero.

PHONES. You can monitor what you are recording through headphones.

TAPE TO TAPE. Press this control to record from one tape machine to another.

PRE-SET STATIONS. You can often pre-set several stations, usually on the VHF waveband, and keep them stored in the tuner's memory. Press the VHF control and turn the tuning knob until the exact frequency you want shows up on the display. Then press one of the memory buttons. By pressing the same memory button again, you recall the station, ready tuned.

QUARTZ SYNTHESIZER. This is a special circuit for very precise tuning.

DIGITAL TUNING DISPLAY. This shows precisely the frequency you are tuned to.

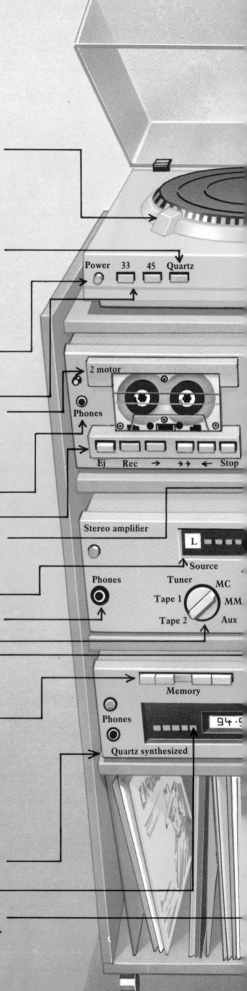

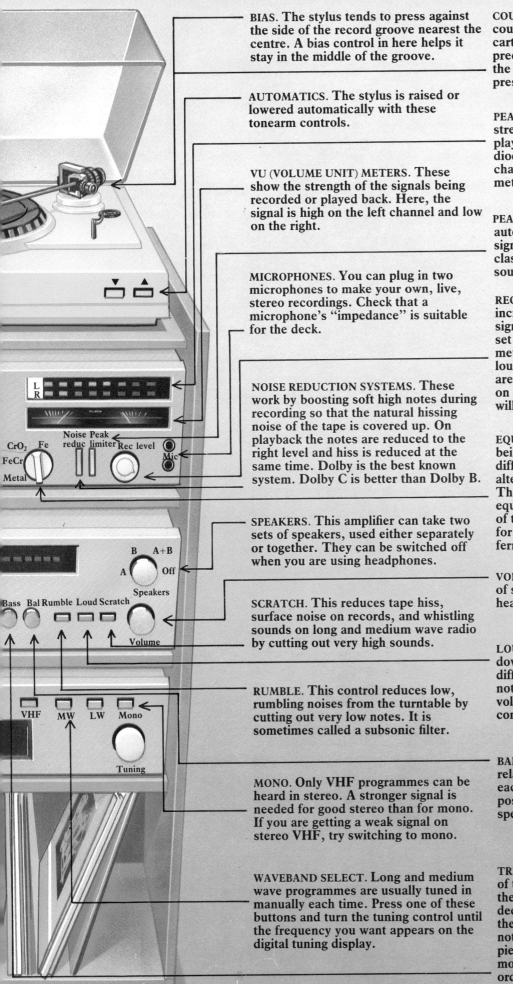

BIAS. The stylus tends to press against the side of the record groove nearest the centre. A bias control in here helps it stay in the middle of the groove.

AUTOMATICS. The stylus is raised or lowered automatically with these tonearm controls.

VU (VOLUME UNIT) METERS. These show the strength of the signals being recorded or played back. Here, the signal is high on the left channel and low on the right.

MICROPHONES. You can plug in two microphones to make your own, live, stereo recordings. Check that a microphone's "impedance" is suitable for the deck.

NOISE REDUCTION SYSTEMS. These work by boosting soft high notes during recording so that the natural hissing noise of the tape is covered up. On playback the notes are reduced to the right level and hiss is reduced at the same time. Dolby is the best known system. Dolby C is better than Dolby B.

SPEAKERS. This amplifier can take two sets of speakers, used either separately or together. They can be switched off when you are using headphones.

SCRATCH. This reduces tape hiss, surface noise on records, and whistling sounds on long and medium wave radio by cutting out very high sounds.

RUMBLE. This control reduces low, rumbling noises from the turntable by cutting out very low notes. It is sometimes called a subsonic filter.

MONO. Only VHF programmes can be heard in stereo. A stronger signal is needed for good stereo than for mono. If you are getting a weak signal on stereo VHF, try switching to mono.

WAVEBAND SELECT. Long and medium wave programmes are usually tuned in manually each time. Press one of these buttons and turn the tuning control until the frequency you want appears on the digital tuning display.

COUNTERBALANCE. This counterbalances the tonearm and cartridge. It has to be positioned very precisely so that the stylus sits lightly on the record with just the right amount of pressure or "tracking force".

PEAK INDICATORS. These show the strength of signals being recorded or played back. They use "light emitting diodes" (LEDs), which respond to changes in signal level faster than VU meters do.

PEAK LIMITER. During recording, this automatically reduces very high level signals, produced, for example, by a clash of cymbals. This helps prevent the sound getting distorted.

RECORDING LEVELS. These controls increase or decrease the strength of the signals being recorded. They should be set so that the needles on the VU meters just reach the red sections in the loudest passages of music. If the levels are too high, the sound will be distorted on playback. If they are too low, you will hear hiss.

EQUALIZATION. When cassettes are being made, the balance between the different pitched notes is artificially altered to get the best results on tape. The balance has to be restored, or equalized, on playback to suit the type of tape. This deck has special circuits for ferro, chrome dioxide (CrO_2), ferro-chrome and metal tapes.

VOLUME. This control alters the amount of signal fed to the speakers or headphones.

LOUDNESS. When the volume is turned down low, say late at night, it can be difficult to hear very high and very low notes. A loudness control boosts the volume of these notes slightly to compensate.

BALANCE. This control affects the relative amounts of sound coming from each speaker. When it is in the central position, equal amounts come from both speakers.

TREBLE AND BASS. You can get the tone of the sounds as you like them with these controls. The treble increases or decreases the volume of the high notes, the bass does the same for the low notes. "Graphic equalizers" are special pieces of equipment which give you more precise control of tone than an ordinary amplifier does.

Make your own radio

These instructions are for building a simple, medium wave radio. You will be able to buy the components you need from an electronics components shop or, if there is no shop near you, by post from an electronics supplier. Look for suppliers' advertisements in electronics hobby magazines. You should be able to buy all the other things from a general electrical or hardware shop.

It is important to follow each step of the instructions very carefully, as the tiniest mistake will probably mean the radio does not work. The number of stations you pick up on your radio will partly depend on where you live. At the bottom of page 26 are some tips on what to do if you live in an area of poor reception.

Components

(You will need one of each of the following.)

Resistors

(1/8 or 1/4 watt, preferably with 5% tolerance, though 2% will do.)

100Ω, 270Ω, 680Ω, 3.9kΩ, 15kΩ, 39kΩ, 100kΩ
(Ω means *ohm*. kΩ means 1,000 *ohms*.)

Variable resistor

25kΩ potentiometer for use as volume control.

Capacitors (wire ended type)

10nF, 100nF (nF means *nanofarad*.)

Electrolytic capacitors

(Axial lead type, 9 volt. If the voltage is any higher, you may find the capacitors are too big for this circuit layout.)

4.7µF, 22µF, 47µF (µF means *microfarad*.)

Variable capacitor

500pF (*picafarad*), single gang, without trimmers, for use as tuner.

Diodes (two of these) IN4148

Transistor BC107

Integrated circuit

ZN414

Audio transformer

With impedance ratio of 1 or 1.2kΩ: 3.2 or 8Ω, e.g LT700.

3 or 8Ω loudspeaker (that is, to match your audio transformer), about 80mm in diameter.

Single pole, on/off, slide switch.

Veroboard with copper strips, 0.1in spacing, at least 30 tracks x 32 holes.

2 metres of enamelled copper wire, s.w.g. 30.

Ferrite rod, 9 or 10mm in diameter, 10cm long.

Other things you will need

Soldering iron, cored solder, wire cutters, sharp knife, small pliers, drill bit (about 4mm), 9 volt battery, sponge, clear sticky tape, needle and thread, 2 or 3 metres of solid conductor insulated "hook-up" wire, thin enough to go through the holes in the Veroboard when stripped. Twin-wire cable can be split in two.

Note. If you are not used to soldering or working with Veroboard, it is a good idea to buy more of th cheapest components (resistors) than you need and an extra piece of Veroboard, so you can practise.

How to solder

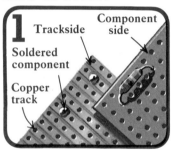

1 Trackside — Component side — Soldered component — Copper track

When components are soldered on to Veroboard, electric current from a battery will flow to them along the copper strips on the back, or trackside, of the board.

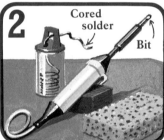

2 Cored solder — Bit

You need a soldering iron, some cored solder and a damp sponge. Plug the iron in and wait for it to heat up. *Prop the iron up carefully so it doesn't burn anything and don't touch the bit.*

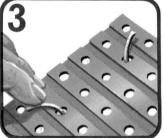

3 Always make sure you put the legs of the components through the right holes in the Veroboard (see next page). Bend them away from each other slightly at the back to hold them in place.

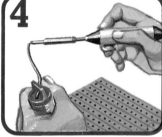

4 Wipe the hot bit on the damp sponge to clean off any old solder. Then, touch the bit with the end of the solder wire. The solder will melt immediately and a drop of solder will cling to the bit.

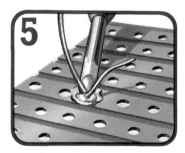

5 Then, both at the same time, put the bit and the solder wire on the track, up against the component's leg. Leave them there for only a few seconds until the solder melts and joins the leg to the track.

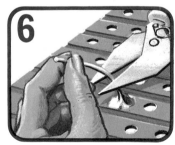

6 Let the "joint" cool and then trim the leg about 1mm from the solder with wire cutters. Hold the leg as you cut to stop it flying into the air. Be careful. Bits of flying metal can be very dangerous.

7 Run iron along groove.

Joints should be smooth and shiny and there must not be any solder between tracks or the current will flow across them. To remove solder, run the hot iron along the groove between the tracks.

8 Remember to unplug the iron when you have finished.

If you solder a component in the wrong place, you can remove it by melting the joint. Put the hot iron on the joint and gradually lift off the solder, wiping the bit on the damp sponge.

24

1 Building the radio

Hold knife firmly against ruler.

If your Veroboard is bigger than you need, you can cut it down. Score along a row of holes on the trackside several times using a sharp knife and metal ruler. Then break the board. Leave a few more tracks and holes than you need in case you don't break the board quite straight.

2

Make a grid to help you find where to put the components. Put the board on a piece of paper component side up, with the tracks running horizontal. Draw round the board. Make marks for the holes along the top and for the tracks down the side. Number the holes 1-32 and the tracks A-DD, as shown above.

3

Broken track

You need to "break tracks" in places. Turn the drill bit in these holes until the copper either side is removed: A14, A18, F26, G12, H14, H20, I18, M21, N11, N21, O21, R22, S11, AA22 and BB22. When you find a hole, try sticking a piece of wire through so you don't lose it again when you turn the board over.

4

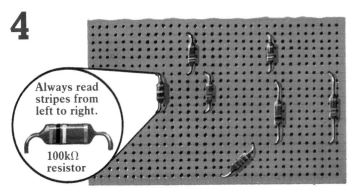

Always read stripes from left to right.

100kΩ resistor

Solder the resistors to the board first. You can tell which is which by their coloured stripes. (Ignore the gold or red stripe round one end.) It doesn't matter which way round you put the resistors. The brown, black, yellow striped resistor (100kΩ) goes in holes I9 and N9. The blue, grey, brown (680Ω) in I15, N15. Orange, white, red (3.9kΩ) in N13, S13. Orange, white, orange (39kΩ) in N23, S23. Red, violet, brown (270Ω) in A17, E21. Brown, green, orange (15kΩ) in E31, N31. Brown, black, brown (100Ω) in E24, M24.

5

10nF capacitor has brown, black, orange stripes from top.

100nF capacitor has brown, black, yellow stripes from top.

Next, put the 10nF capacitor in J6 and N6, and the 100nF capacitor in E13 and I13. 10nF is sometimes printed on capacitors as 0.01μF and 100nF as 0.1μF. Or, your capacitors may have coloured stripes like the ones shown in the diagram above.

6

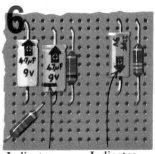

Indicates negative end   Indicates positive end

Electrolytic capacitors have to go a certain way round. Put the 22μF capacitor in F29, N29 with the positive end at F. The 47μF goes in E22, M22 with the positive end at M. The 4.7μF goes in E19, N19 with the positive end at N.

7

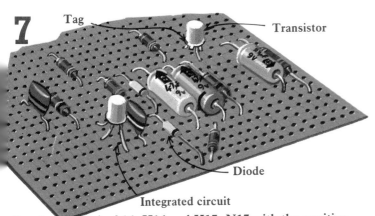

Tag   Transistor

Diode

Integrated circuit

Put the diodes in A16, H16 and H17, N17 with the positive ends (marked with a ring) towards A and H. Don't overheat the diodes, transistor or integrated circuit or you may damage them. The transistor goes in M26, N27, O26, with the leg nearest the metal tag in M26. The integrated circuit goes in E11, G10, I11, with the leg nearest the tag in I11. Leave both the transistor and integrated circuit sticking up about 1cm from the board.

8

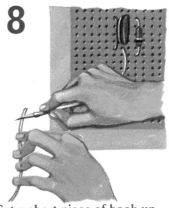

Cut a short piece of hook-up wire. Using a sharp knife, carefully strip about 1cm of the plastic casing off each end. Put one end through hole E7, the other through J7 and solder. Solder another short wire link from O28 to R28.

9

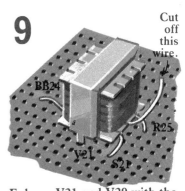

Cut off this wire

BB24

R25

V21   S21

Enlarge V21 and V29 with the drill bit, so the transformer's legs will go through. Put the wires in R25, S21, AA26 and BB24, as shown above. Make sure the transformer is the right way round. Cut off the extra wire. Bend the legs flat on the trackside. Solder them and the wires.

25

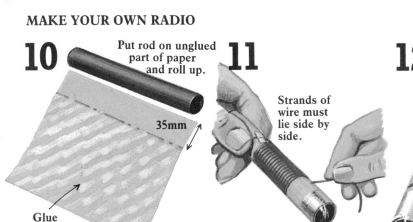

10 Put rod on unglued part of paper and roll up.

35mm

Glue

11 Strands of wire must lie side by side.

12

13

Now make the aerial. If your ferrite rod is too long, cut a groove round it with a hack-saw, 10cm from one end. Then break it. Make a tube for the rod out of a thick piece of paper, as shown above. Put sticky tape round the ends of the tube, so the rod doesn't fall out.

Starting 4cm from one end of the tube and leaving about 5cm of wire free, wind enamelled copper wire neatly round the tube 36 times. Tape the end of the wire to the tube to hold it in place as you wind. Leave 5cm free at the other end. Cover all the wire with clear sticky tape.

Scrape both ends of the wire with a knife to remove the coating. Touch the soldering iron with solder and quickly stroke the ends to give them a coating of solder. Then solder them in G5 and N5. Don't stretch the wires tight. Fasten the aerial to the board with a needle and thread.

Cut nine pieces of hook-up wire about 15cm long. Strip the casing off the ends and solder one end of each piece in these holes: E14, E30, F28, G1, I14, N1, S31, AA30, BB30

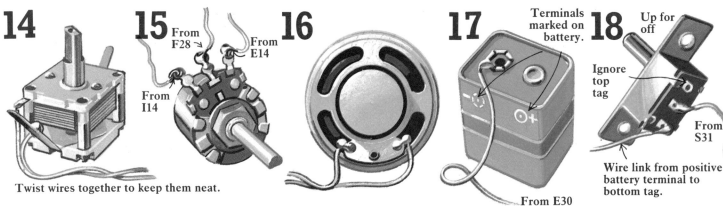

14

Twist wires together to keep them neat.

15 From F28   From E14

From I14

16

17 Terminals marked on battery.

– +

From E30

18 Up for off

Ignore top tag

From S31

Wire link from positive battery terminal to bottom tag.

Solder the wire from N1 to one of the long legs on the tuner and the wire from G1 to the short leg.

Solder the wires from E14, I14 and F28 to the volume control as shown above. Hook wires through holes to solder.

Solder the wires from AA30 and BB30 to the speaker tags. It does not matter which way round they go.

Solder the wire from E30 to the negative battery terminal. Make quite sure you have the correct terminal.

Cut another length of hook-up wire and strip the ends. Connect the switch to the board and battery as above.

Checking the radio

Before you switch on, check all the components on the board to make sure they are the right ones, that they are in the right holes and are the right way round. Check you have broken tracks in the right places and that the wires to the other components are connected correctly. Check again that the battery is the right way round. If you get it wrong, not only will you have to correct it, you may have to replace the transistor, diodes and integrated circuit with new ones, as they could be ruined.

Now, turn up the volume and switch on. Turn the tuner slowly up and down the waveband to see how many stations you can get. If the sound is not very good, try moving the radio around to get better reception.

If the radio doesn't work, switch off

and check everything again. Examine all the joints very carefully to make sure they are firm. Try waggling the components or giving them a tug. Make sure there is no solder between tracks and that the components' legs are not touching each other on the component side of the board. Try another battery, in case yours is flat.

If you still can't find anything wrong, ask someone else to check the board for you. They may spot something you didn't. If, after all that, you still can't get the radio working, send it to us and we will try and find out what is wrong. Wrap it up very carefully and send it (with enough stamps to pay for the return postage) to: Electronics Adviser, Usborne Publishing Ltd, 20 Garrick Street, London WC2E 9BJ, England.

Expert's tips

If you live near a strong transmitter, you may only pick up that station. If this happens, take out the resistor at A17, E21 and replace it with a piece of wire.

If you live in an area of weak reception, try rigging up an extra aerial with a few metres of wire. First, put an extra 10nF capacitor at C6, G6 and break the track at C8. Then, solder one end of the aerial wire in C1 and put the other end somewhere high up, for example, over a curtain rail.

If you want, you can box your radio. Make holes for the switch, speaker, and volume and tuning controls, and mount them in position, re-soldering as necessary. Keep the tuner and its wires away from the others. Pad the radio well.

CB and two-way radio

One of the most common uses of radio is for two-way communication. This is especially useful for groups of people who need to communicate while on the move, such as the fire brigade, police, ambulance drivers, pilots, ships' radio officers, taxi drivers and astronauts. Different radio frequencies are allocated to the different users of two-way radio, usually on the VHF or UHF (ultra-high frequency) wavebands. The type of transmitters and receivers used depends on the waveband and the distance the messages have to travel. A "transceiver" is a transmitter and receiver in one unit.

Citizens' band

Anyone can become a "CBer" and talk to other people by radio. All you need is a licence from the Post Office and a transceiver, or rig. The first CB rigs were used in lorries or cars but you can also buy base rigs to use at home, or small pocket-size sets. They are all easy to operate. The signals will travel only a short distance in built-up or hilly areas and up to about 30km in open country. CBers are supposed to keep their conversations short to stop their radio channels getting jammed and they often talk in a number code. If you want to find out more, look in the CB magazines to find out where your nearest club is.

Amateur radio

If you are interested in radio technology, you might like to become an amateur radio "ham" and talk to people all over the world. To do this, you have to be at least 14, hold a licence and pass exams to show you understand the equipment and know how to operate it. A lot of amateurs build at least part of their equipment themselves. It is possible to transmit from a car as well as from home. As well as talking normally, amateurs communicate in Morse code and, for exchanging technical information, in an internationally understood code of three letters. Each radio ham has a "call sign" of letters and numbers, which identifies the individual and the country they are speaking from. They often send "QSL" cards to each other, confirming their conversations. It is illegal to use amateur radio for business or propaganda purposes. To find out more, write to your national radio society, enclosing an s.a.e. They will tell you where the nearest local branch is. (The address in Britain is: Radio Society of Great Britain, 35 Doughty Street, London WC1N 2AE.)

If you want to listen in to amateurs' conversations but don't want to make transmissions, you can use a high quality short wave receiver. You do not need a licence just for listening.

Car telephones

Conversations go from the car to the telephone exchange and back again by radio. The link from the exchange to the other person in the call is made on the ordinary telephone network.

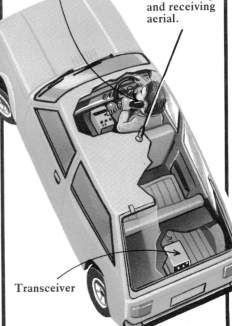

Telephone

Transmitting and receiving aerial.

Transceiver

Different uses of radio and audio

Radio and, less often, audio are used in many ways in addition to the ones explained so far in this book. As well as carrying sound, pulses of radio waves can be used to control things and to find out information about things. Here are some examples of the different uses.

Television

Both the pictures and sound for television programmes are transmitted on radio waves. The TV cameras convert light into electronic signals for the pictures. Then, both these and the sound signals are broadcast on the VHF or UHF wavebands.

Dish aerial

Receiver

Radio waves hit dish and are reflected on to receiver.

Aerial can be steered to face in different directions.

Astronomy

Many objects in space give out radio waves instead of, or as well as, light. Astronomers can study them using radio telescopes like the ones on the left. The telescopes act like giant radio receivers. They pick up the signals from space, which can then be recorded on tape and fed to computers for analysis.

Radar

Ship's radar scanner rotates.

Radar screen

Lighthouse

Land

Other ship

Radar is most often used to work out the positions of ships and aircraft. A beam of radio waves is sent out from a scanner, usually a rotating one.

When the waves hit objects, especially metallic ones, some of their energy bounces back off the objects. From these "echoes" it is possible to plot a picture of the surrounding area and work out how far away things are. Radar is short for "radio detection and ranging".

Radio microphones

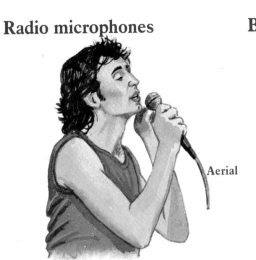

Aerial

These are sometimes used at concerts or in TV studios to avoid having lots of trailing wires. The microphone has a transmitter inside it. Receivers are set up out of sight. They pick up the transmitted sound and feed it either to loudspeakers or to the control desk.

Bleepers

Radio receiver

Some people, hospital doctors for instance, carry small radio receivers around in their pockets. When they hear a "bleep" on the radio, they know they have to go to, or should telephone, a certain place. This system is called radio-paging.

Model control

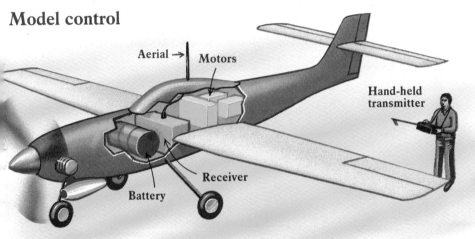

Aerial → Motors

Hand-held transmitter

Receiver

Battery

Model planes and boats can be controlled by radio. The operator uses a small transmitter to send out coded radio signals telling the model how to move. The model has a receiver in it. This interprets the coded instructions and passes them on to the motors which work the different parts of the model.

Sonar

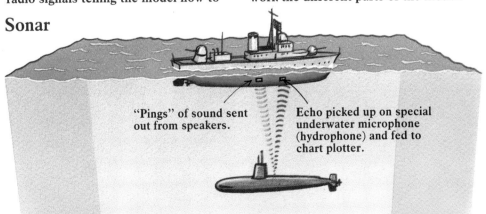

"Pings" of sound sent out from speakers.

Echo picked up on special underwater microphone (hydrophone) and fed to chart plotter.

Most radio waves do not travel well through water but sound waves do. Ships can find out if there are submarines nearby by sending out bursts of sound. If the sound hits a submarine, an echo comes back. The position of the submarine and its distance away is then automatically plotted on a chart. The same method is used to find out how far down the sea-bed is, and to locate shoals of fish. Sonar is short for "sound navigation and ranging".

Weather balloons

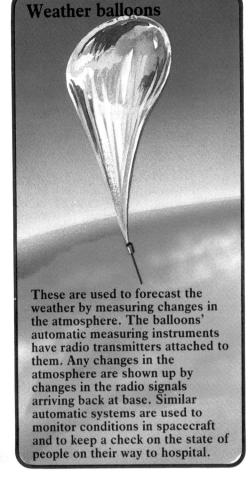

These are used to forecast the weather by measuring changes in the atmosphere. The balloons' automatic measuring instruments have radio transmitters attached to them. Any changes in the atmosphere are shown up by changes in the radio signals arriving back at base. Similar automatic systems are used to monitor conditions in spacecraft and to keep a check on the state of people on their way to hospital.

Plane on right course.

Aircraft landing

Radio is very important in helping planes to navigate. Coded radio signals are sent out automatically from airports. These tell pilots whether they are on course, first for the airport, then for the runway. One signal tells them whether they are right or left of the runway, another whether they are too high or too low.

The future

Experts are always trying to find new ways of improving the quality of sound recording and reproduction and to develop new techniques for radio broadcasting. Two of the most revolutionary advances are, in audio, the invention of the compact, digital disc and, in radio, the use of satellites. These are described on the right, along with some of the other things you can expect to find in a home of the future.

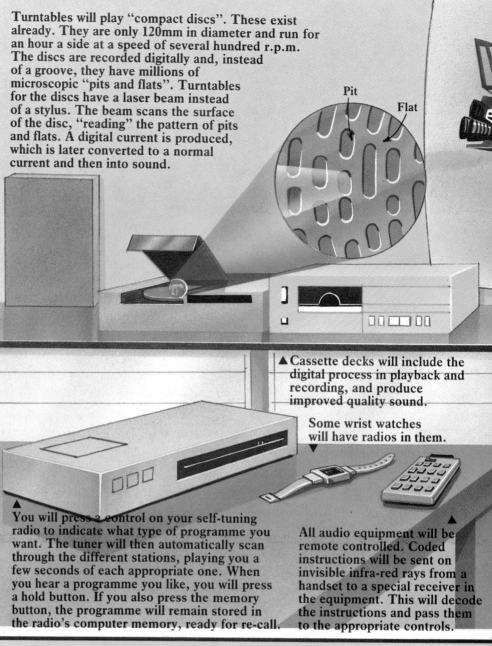

Turntables will play "compact discs". These exist already. They are only 120mm in diameter and run for an hour a side at a speed of several hundred r.p.m. The discs are recorded digitally and, instead of a groove, they have millions of microscopic "pits and flats". Turntables for the discs have a laser beam instead of a stylus. The beam scans the surface of the disc, "reading" the pattern of pits and flats. A digital current is produced, which is later converted to a normal current and then into sound.

Pit

Flat

▲ Cassette decks will include the digital process in playback and recording, and produce improved quality sound.

Some wrist watches will have radios in them.

▲ You will press a control on your self-tuning radio to indicate what type of programme you want. The tuner will then automatically scan through the different stations, playing you a few seconds of each appropriate one. When you hear a programme you like, you will press a hold button. If you also press the memory button, the programme will remain stored in the radio's computer memory, ready for re-call.

All audio equipment will be remote controlled. Coded instructions will be sent on invisible infra-red rays from a handset to a special receiver in the equipment. This will decode the instructions and pass them to the appropriate controls.

Digital sound

Processing sound digitally (by numbers) cuts out distortion almost completely. The process can already be used in making tapes and records and will also be used for radio and TV sound. The electric signals from a microphone are scanned several thousand times a second and are given a number depending on their strength at each moment of scanning. The string of numbers, in the form of electric pulses, can travel by wires or radio, even over great distances, without the quality of the final sound suffering at all. This digital current has to be converted back into the normal type of current, called an analogue current, before being fed to loudspeakers.

Audio and radio words

Here are some audio and radio words and their meanings. If the word you want is not shown in this list, look it up in the index, as it may be mentioned elsewhere in the book.

ANALOGUE. An electric current which flows smoothly, unlike a digital current, which is in pulses.
ANTENNA. Aerial.
AUTOREVERSE. Cassette deck which plays or records both sides of a tape without you having to turn the cassette over.

BANDWIDTH. The range of sound frequencies that a piece of equipment can handle. Measured in hertz. The wider the range the better. Don't confuse bandwidth with waveband.
BIAS. A special signal used in tape recording to avoid distortion. Some cassette decks have a switch for you to select the right bias level for the type of tape you are using, some do it automatically. Also, a control in turntable tonearms which helps the stylus sit centrally in the record groove.
CROSSOVER POINT. The particular frequencies at which signals to a loudspeaker are divided up to be handled by the different sorts of speaker (tweeters, squawkers, woofers).

DECIBEL (dB). A unit of measurement of the loudness of sounds, or the size of electric signals.
DIGITAL. An electric current which is in a series of pulses.
DIGITAL DISPLAY. A display given in numbers, as on the frequency display of tuners and receivers.
DIN. A German organization which lays down internationally recognized standards for audio equipment. Anything which meets the DIN standard 45500 qualifies as hi-fi.
DISTORTION. The difference between the signal which goes into an amplifier and the one that comes out. Measured as a percentage. Look for a "total harmonic distortion" (THD) figure which is as low as possible.
DUAL CAPSTAN. Cassette deck with two

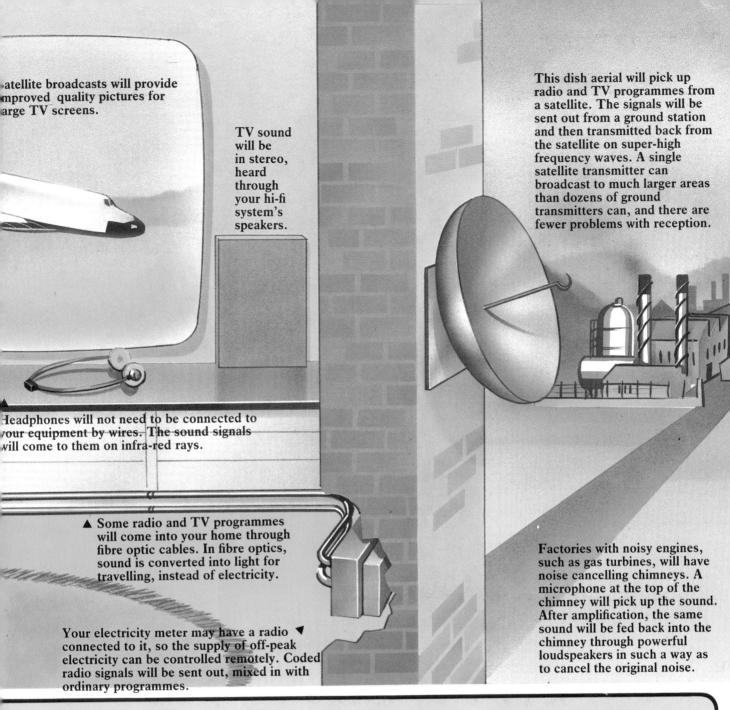

satellite broadcasts will provide improved quality pictures for large TV screens.

TV sound will be in stereo, heard through your hi-fi system's speakers.

This dish aerial will pick up radio and TV programmes from a satellite. The signals will be sent out from a ground station and then transmitted back from the satellite on super-high frequency waves. A single satellite transmitter can broadcast to much larger areas than dozens of ground transmitters can, and there are fewer problems with reception.

Headphones will not need to be connected to your equipment by wires. The sound signals will come to them on infra-red rays.

▲ Some radio and TV programmes will come into your home through fibre optic cables. In fibre optics, sound is converted into light for travelling, instead of electricity.

Factories with noisy engines, such as gas turbines, will have noise cancelling chimneys. A microphone at the top of the chimney will pick up the sound. After amplification, the same sound will be fed back into the chimney through powerful loudspeakers in such a way as to cancel the original noise.

Your electricity meter may have a radio ▼ connected to it, so the supply of off-peak electricity can be controlled remotely. Coded radio signals will be sent out, mixed in with ordinary programmes.

capstans for pulling the tape past the heads. This helps keep the tape speed constant.
FREQUENCY RESPONSE. Same as bandwidth. See opposite page.
GIGAHERTZ (GHz). 1GHz = 1,000MHz.
HERTZ (Hz). The unit of measurement of frequency. 1,000Hz = 1kHz.
HF. High frequency. The high frequency waveband ranges from 3 to 30MHz.
IMPEDANCE. The resistance of certain components in a circuit to the flow of electricity. Measured in ohms. Loudspeakers and headphones must have the correct impedance for the system's amplifier to prevent distortion of sound or damage to the equipment.
KILOHERTZ (kHz). 1kHz = 1,000Hz.
LF. Low frequency. The LF waveband ranges from 30 to 300kHz.

LOW MASS. Lightweight tonearm and cartridge.
MEGAHERTZ (MHz). 1MHz = 1,000kHz.
METAL COMPATIBLE. Cassette deck which will play metal type tapes.
MF. Medium frequency. The MF waveband ranges from 300kHz to 3MHz.
MONO. One-track sound, that is with only one speaker or microphone.
PHONO. Label on amplifier switch which selects signals from turntable.
PICKUP ARM. Tonearm.
SENSITIVITY. The amount of signal which has to be fed into a piece of equipment in order to produce a given amount of power in the loudspeaker.
SEPARATION. The amount by which the signals from one channel of, say, a stereo system will "leak" on to the

other, producing a slight mixture in the speakers. Measured in dB. The higher the figure the better.
SHF. Super-high frequency. SHF radio waves range from 3 to 30GHz.
SIGNAL TO NOISE RATIO. The size of the wanted signal compared with unwanted noise produced in the equipment's circuits. Measured in dB. The higher the figure the better.
STEREO. Two-track sound system, with two microphones and two loudspeakers.
UHF. Ultra-high frequency. UHF radio waves range from 300MHz to 3GHz.
VHF. Very high frequency. The VHF waveband ranges from 30 to 300MHz.
WAVEBAND. A particular range of radio frequencies, e.g. LF, VHF.

Index

aerial, 3, 5, 7, 8, 10, 11, 12, 13, 19, 26, 27, 28, 30, 31
   dish, 28, 31
aircraft, 28, 29
amateur radio, 27
amplifiers, 11, 18, 19, 21, 22, 23, 30, 31
   integrated, 21
   power, 21
amplitude, 8, 10, 11
   modulation, 10, 11
analogue current, 30
astronomy, 28

balance, 21, 23
bandwidth, 30, 31
bass, 23
battery, 8, 9, 12, 13, 18, 19, 24, 26, 29
bias, 23, 30
bleepers, 29

capacitors, 12, 24, 25, 26
capstan, 20, 21, 22, 30
car audio, 19
cartridges, 20, 23, 31
   moving coil, 20, 22
   moving magnet, 20, 22
cassette decks, 19, 20, 21, 22, 23, 30, 31
   recorders, 17, 19
cassettes, 7, 19, 21, 23
   making of, 14-17
CB, 27
chrome dioxide (CrO_2) tape, 21, 23
compact discs, 30
computers, 28, 30
continuity announcer, 4
   desk, 4
control desk, 4, 6-7, 14-15, 16, 29
   room, 4, 6-7, 14-15
counterbalance, 23
cutting head, 17

delay-box, 7
demodulation, 13
diamond stylus, 17, 20
diaphragm, 3
digital,
   current, 30
   delay, 16
   display, 22, 23
   records, 21, 30
   sound, 30
diodes, 12, 13, 24, 25, 26
   detector, 13
   light emitting (LEDs), 23
disc jockey, 6-7
distortion, 22, 23, 30, 31
Dolby, 23

Earth, 10
echo, 15, 16
editing, 6, 14, 15, 21
electromagnetic spectrum, 9
engineer, 4, 6-7, 15, 16
EPs, 22
equalization, 23
erase head, 15, 20, 21
erasing, 15, 21

fader, 6, 15
ferro tape, 23
ferro-chrome tape, 23
fibre optics, 31
flutter, 22
frequency, 3, 8, 10-11, 13, 15, 22, 23, 27, 31
   modulation, 10, 11

gamma rays, 9
gigahertz (GHz), 31
graphic equalizers, 23

headphones, 2, 4, 7, 14, 15, 18, 19, 20, 22, 23, 31
hertz (Hz), 30, 31

Hertz, Heinrich, 9, 10
hi-fi, 2, 18, 19, 30, 31
   mini, 18, 19
hydrophone, 29

impedance, 20, 22, 23, 24, 31
infra-red, 9, 30, 31
input, 22
integrated circuit, 24, 25, 26
interference, 9, 11
interviews, 6, 7
ionosphere, 10, 11
iron oxide, 14, 15

jackfield, 7
jingles, 4, 6, 7

kilohertz (kHz), 10-11, 31

laser beam, 30
light, 9, 28, 31
linear tracking, 20
loudness control, 23
loudspeakers, 2, 4, 5, 7, 12, 13, 15, 16, 19, 20, 21, 24, 26, 29, 30, 31
LPs, 22

magnet, 3, 14, 17, 20, 21
magnetism, 3, 15
megahertz (MHz), 11, 31
memory, 22, 30
metal tape, 21, 23, 31
microphones, 3, 4, 6, 7, 11, 14, 15, 19, 23, 29, 30, 31
   capacitor, 21
   cardioid, 21
   condenser, 21
   dynamic, 21
   moving coil, 3, 21
mixer, 13, 14-15, 16
mixing down, 16
model control, 29
modulation, 10-11
modulator, 11
modules, 15, 16
mono, 18, 19, 23, 31
music centres, 18

navigation, 28, 29
news bulletins, 6
noise cancellation, 31
   reduction, 23

ohms, 24, 31
oscillator, 11, 13
overdubbing, 15

pan pot, 15, 16
peak indicators, 23
   limiter, 23
phase-locked loop (PLL), 22
phone-ins, 7
pinch wheel, 20
pitch, 3, 20
playback head, 20, 21
playing back, 21, 22, 23
power indicators, 22
   output, 22
preamplifier, 21
pre-setting, 22
producer, 6, 14-15
production secretary, 6, 7
programmes, 4, 6-7, 19, 23, 30, 31

'QSL' cards, 27
quartz crystal, 11
   lock, 22
   synthesizer, 22
rack systems, 18, 22-23
radios, 2, 5, 12-13, 18, 19, 24-26, 30, 31
radio, alarm clocks, 19
   cars, 7
   /cassette recorders, 18

'ham', 27
paging, 29
societies, 27
station, 4, 10, 22, 24, 26, 30
studio, 4, 6-7
two-way, 7, 27
receiver, 3, 12, 27, 28, 29, 30
receivers, 19
reception 11, 24, 26, 31
recording, 14-15, 19, 22, 23, 30
   head, 14, 20
   levels, 23
   studios, 14-15
record press, 17
records, 6, 7, 21, 30
   making of, 14-17
reel-to-reel recorders, 4, 6, 14, 15, 16, 21, 22
relay transmitters, 5
remote control, 30
   studios, 7
reporter, 6, 7
resistors, 12, 24, 25, 26
r.p.m., 22, 30
rumble, 20, 23

sapphire stylus, 20
satellite broadcasting, 11, 30, 31
scratch, 23
SHF, 11, 31
ships, 27, 28, 29
signal strength indicator, 22
soldering, 24
soloing, 15
sonar, 29
sound effects, 7, 16
   signals, 3, 4, 5, 7, 10, 11, 12, 13, 14, 15, 16, 21, 22, 23, 28, 30
   waves, 2, 9, 21, 29
source, 21, 22
space, 8, 11, 28
spacecraft, 29
speed, 21, 22, 30, 31
spools, 14, 15, 22
squawkers, 20, 30
stampers, 17
stars, 8
stereo, 16, 17, 18, 19, 21, 23, 31
stroboscopes, 22
stylus, 17, 20, 23, 30
subsonic filter, 23
surround-sound, 19

tape, 14-15, 20, 21, 23, 28
   counter, 22
   hiss, 23

master, 16, 17
   recorders, 4, 6, 14, 15, 16, 21, 22
   types of, 21, 23
tapes, 6, 16, 21, 30
telephone, 3, 4, 7
   car, 27
telescopes, 28
television, 9, 21, 28, 29, 30, 31
time-signals, 4
tone, 4, 7, 15, 16, 21, 23
tonearms, 20, 23, 30, 31
track, 14, 15, 16, 17, 21
tracking force, 23
transceiver, 27
transformers, 12, 24, 25
transistors, 12, 21, 24, 25, 26
transmitter, 3, 4, 5, 7, 8, 10, 11, 12, 13, 24, 26, 27, 29, 31
transmitting, 5, 10-11
treble, 23
tuner, 13, 24, 26, 30
tuners, 18, 19, 22
tuning, 5, 10, 13, 22, 30
turntables, 7, 17, 18, 19, 20, 22, 23, 30, 31
   belt driven, 20
   direct drive, 20
   idler drive, 20
tweeters, 20, 30

UHF, 10, 27, 28, 31
ultra-violet, 9

Veroboard, 24, 25, 26
VHF, 5, 10, 12, 13, 19, 22, 23, 27, 28, 31
voice-overs, 6
volume, 4, 6, 7, 12, 13, 15, 23, 24, 26
VU meters, 4, 15, 23

waveband, 8, 10-11, 12, 13, 18, 19, 22, 23, 27, 28, 30, 31
wavelength, 8, 9, 10-11,
waves,
   carrier, 5, 8, 10, 11, 12, 13
   electromagnetic, 8
   long, 5, 8, 10, 11, 12, 13, 19, 23
   medium, 5, 8, 10, 11, 12, 13, 19, 23, 24
   radio, 3, 5, 8-9, 10-11, 28, 29
   short, 10, 27
   sound, 2, 9, 21, 29
weather balloons, 29
woofers, 20, 30
wow, 22

X-rays, 9

First published in 1982 by Usborne Publishing Ltd, 20 Garrick Street, London WC2E 9BJ, England.